Lasers and Holograms

LASERS AND HOLOGRAMS

JUDY ALLEN

Illustrated by Andrew Skilleter
Consultant: Chris McDowell

PEPPER PRESS, 1983

Consultant: Chris McDowell

With special thanks to Optec
Design and Rob Rimmington for
the hologram.
Hologram courtesy of Light Impressions.
Hologram Copyright © Light Impressions.

Printed and bound in Great Britain by
R. J. Acford Ltd, Chichester
for Pepper Press, 1983

An imprint of
Bell & Hyman Limited,
Denmark House,
37–39 Queen Elizabeth Street,
London SE1 2QB.

ISBN: 0 237 45668 0

The laser is a machine designed to work with light, which is why it can create such stunning visual effects. It can manipulate visible light so that it moves in vividly-glowing criss-crossing colours for a stage show, or so that it creates those ghostly three-dimensional pictures known as holograms. On a rather less domestic level, it can bounce a beam of light off the moon, a quarter of a million miles away, or create real (if tiny) stars in a laboratory. It is so straight that it can measure lengths and speeds with almost perfect precision. It is so accurate that it can be used to perform the most delicate of eye operations.

But light is not only something that can be seen, it is also a form of energy, and it is the cutting and burning energy of the laser light that makes it one of the most dramatic of modern inventions. The beam is so powerful that it can cut through metal, and even the hardness of diamonds, quickly and cleanly. All its science-fiction attributes – knocking out enemy spacecraft or threatening to separate the hero's torso from his legs – are, in fact, possible.

The word itself is an acronym – being made up from the initials of the words which explain it. So LASER comes from the phrase Light Amplification by Stimulated Emissions of Radiation. (This may or may not explain everything. If not, read on.)

In fact, what happens inside the laser to enable it to give out its uniquely powerful beam of light is not only Amplification (enlarging or strengthening the light) but also Oscillation (the shaking or vibrating of the particles which actually produce the light). In its early days the laser came within a hair's breadth of being described as Light Oscillation by Stimulated Emissions of Radiation, and therefore being known as a LOSER. However, it was felt that those who provide money for the development of new scientific wonders would be unlikely to want to back a loser, so laser it is.

The achievements of the laser beam are only possible because the laser itself amplifies light. Its power is further increased because the light is prevented from scattering and is forced to flow uniformly in one direction. So in order to understand how the laser works, it is helpful first to understand something about light itself . . .

Beyond the Rainbow

The part of the spectrum that most people are aware of is visible light. White light can be broken down into an endless number of colours – ranging from violet, through indigo, blue, green, to yellow, to orange and red – the colours that always appear (and always in the same order) in a rainbow.

But visible light is only a tiny part of a much larger spectrum of electro-magnetic radiation (see diagram). The term 'electro-magnetic' is used for radiation, or waves, which travel at the speed of light and which have different properties according to their different wavelengths or frequencies.

Wavelengths and Frequencies

Every part of the whole electro-magnetic spectrum vibrates at a different frequency, from the very high-frequency X-rays to the far lower frequency radio waves. (Though even the lower frequencies are very high and electro-magnetic waves measure

in at about 55 millionths of a centimetre). There may well be frequencies above and below those shown which have not been detected yet and whose properties are still unknown.

Each electro-magnetic radiation can be thought of as a 'train' of waves, and these can be measured in two ways. The length between each wave can be measured to give, quite literally, the wavelength. And the frequency, or number of wave crests passing per second, can also be measured.

The diagram shows that the more frequent the wave crests, the shorter the waves. So – the higher the frequency, the shorter the wavelength, and the lower the frequency the longer the wavelength.

From Ultra-Violet to Infra-Red

In the spectrum of visible light the highest frequency we can comfortably see is the one which produces violet. Beyond that lies ultra-violet which is invisible to the eye – although it can produce

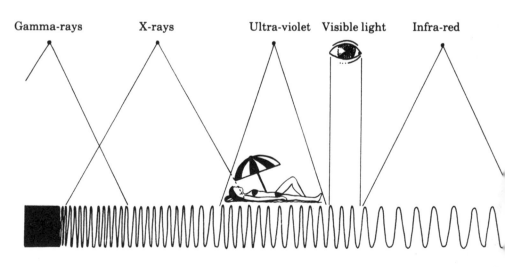

Gamma-rays   X-rays   Ultra-violet Visible light   Infra-red

visible effects, of which sun-tanned skin and sun-bleached hair are two of the most obvious. The lowest frequency we can easily see is the one which produces red, although the eye can sometimes pick up a dim glow from the infra-red frequencies below it.

Light As Energy

Light is not only something that can be seen, it is also energy. Energy cannot be created out of nothing, but one kind of energy can be changed into another. So electric energy can be converted into light energy in an electric light bulb and heat energy can be converted into light energy in a candle flame. Equally, light energy can become heat energy.

The Infra-Red Hot Poker

The higher the frequency of light, the greater its energy. In other words, blue light has more energy than red. Watch a poker heating up in a coal fire. As it heats, it gives off heat energy which can be seen as light. It starts with the low infra-red

rays, which can't be seen clearly but which feel hot. As the heat energy increases, the particles of matter which make up the poker vibrate at higher frequencies, from red to orange and yellow. The poker will eventually become white-hot.

The reason the poker does not work its way up to a pure violet, as might be expected, is because even as the particles within it begin to vibrate at higher frequencies, they continue to vibrate at the lower frequencies as well. By the time the domestic poker is as hot as it can ever be, it is vibrating on all visible frequencies at the same time – and the mixture of all these frequencies results in light of a non-colour, called white. There is a limit to the heat a domestic fire can produce, so there is a limit to the temperature the poker can reach. It would never be able to go to the extreme of moving up out of the visible spectrum and giving off ultra-violet rays – but the hottest of the stars, including our sun, do just that.

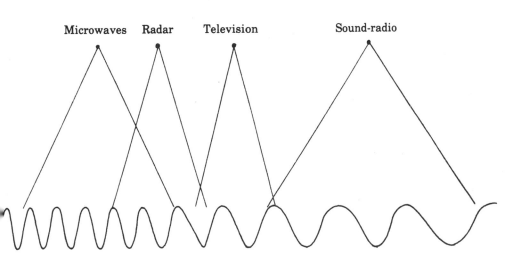

Microwaves   Radar   Television   Sound-radio

The actual production of light takes place on an extremely small scale, far below anything that can possibly be seen, down among the atoms and their electrons.

An atom is the smallest particle of any element that cannot be divided without being destroyed – but it is not solid. It is often represented (in a very simplified way) as being made up of a nucleus, or centre, of protons and neutrons, surrounded by an orbiting cloud of electrons.

These electrons behave in specific ways – and exist in specific energy states – which science can measure. They have, in fact, four principal attributes. The first is the electron's distance from the centre of the atom, its orbit. The second is the ellipticity, or shape, of that orbit. The third is the direction of spin of the electron in relation to the direction of its rotation. The fourth is its magnetic value.

What is more – there are maximum numbers of electrons allowed in any one orbit – for example, 2 in the first orbit, 8 in the second, 18 in the third and 32 in the fourth. And no two electrons in any one atom can have all four principal attributes the same. That is – while two or more electrons can share an orbit, for example, the ellipticity of that orbit, the spin and the magnetic value of each electron would be different.

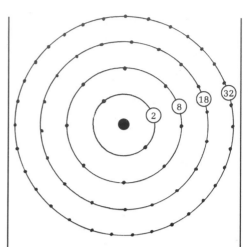

There are maximum numbers of electrons allowed in any one orbit

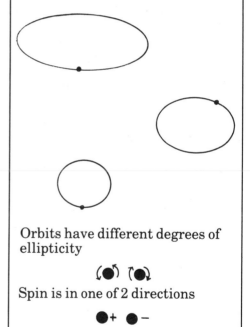

Orbits have different degrees of ellipticity

Spin is in one of 2 directions

There are different magnetic values

Given all the possible combinations of the four attributes, it is obvious that electrons in any one atom exist in a variety of energy states. Yet, although so much is happening within it, this is a description of a stable atom – and the electrons in a stable atom don't give out energy. If light energy is to come out, the atom must first be made unstable, and this is done by feeding some kind of energy in from outside.

When an atom is given an extra shot of energy – electrical energy, perhaps – its electrons absorb this energy and begin to orbit faster. Because of this extra speed some of them will rise into a higher energy orbit – and then jump back into a lower energy orbit, giving off electro-magnetic waves as they do so.

Each electron must rise one or more complete orbits and jump back one or more complete orbits, or nothing will happen.

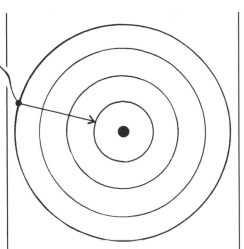

It will then fall back to its original place, or an intermediate level, giving out electro-magnetic radiation as it does so.

These movements of the electron, to a higher energy orbit and then back to a lower energy one, are known as transitions. And the amount of energy the electron absorbs, and then gives off as it falls back, is determined by the transition it has made. If it has made a large transition, between orbits of very high and low energy, it will give out more energy than if the transition was a lesser one.

It is the electron transitions that define the frequency and wavelength of the energy given off by each single falling electron. So, if the electron transition is between the right levels, visible light can be produced.

This electron can jump

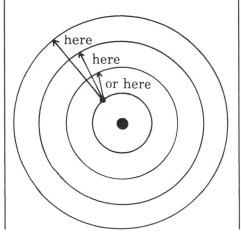

here
here
or here

The Particle Theory versus The Wave Theory

Light can be thought of as waves or as streams of almost indescribably minute particles of matter. During the 17th and 18th centuries science argued the rival merits of the particle theory and the wave theory – and, in fact, the discussion still raises its head from time to time.

The main argument against the particle theory was that experiments had already shown that light usually behaves as a wave. The main argument against the wave theory was that waves are only a disturbance of something else (sea waves are a disturbance of water and sound waves are disturbances of the air) and no one had succeeded in discovering what it was that the light waves were disturbing.

The happy solution has been to think of light as particles of energy which move in a wave-like manner. When experiments are conducted into the behaviour of light it is sometimes regarded as waves and sometimes as particles depending on which is most useful for the experiment. The smallest possible particle of light energy is called a photon.

How Ordinary Light Behaves

But, as has already been seen, light has many different possible frequencies and wavelengths – the whole spectrum of visible light (contained within the larger spectrum stretching before it and beyond it). And ordinary light is usually vibrating on several wavelengths and frequencies at the same time. What is more, the waves are not in step with each other. They can reinforce each other or cancel each other out; just as two sea waves together can build into a larger wave, while two waves which are out of sequence (so that a peak and a trough arrive together) can result in a very feeble wave.

Consider the beam from an ordinary torch. Use the particle theory, thinking of the light as a stream of minute photons. Then think of the stream of photons as a stream of people, wandering along more or less at random. These people are of different heights and weights and lengths of leg, and so they have different lengths of stride (or different wavelengths). Their feet strike the ground in different rhythms (or at different frequencies). What is more, they don't all walk in the same direction but diverge outwards so that their energy is gradually dispersed.

The light from an ordinary battery torch may not be able to travel farther than the end of the garden. Laser light, on the other hand, can reach the moon. So what does a laser do to light?

The laser amplifies light by stimulating electrons to give off electro-magnetic radiations in some parts of the spectrum, in other words, to create photons.

In normal circumstances a particle of light, a photon, is created because an intake of energy causes the electrons orbiting an atom to move to a higher energy orbit from which they may then drop back (see page 11).

So the laser must contain a collection of atoms to work with. This is known as the lasing medium and can be in the form of a gas, a liquid, or a solid crystal. The laser must also have the means of pumping in the energy to cause transitions of electrons from one orbit to another.

When energy is forced into the laser material, some electrons are 'pumped-up' to higher energy levels, which are generally unstable, and consequently these electrons will fall back down to a lower energy level, and photons of light are produced. The light energy does not disperse because the atoms are enclosed within the laser tube or resonator. This is blocked at each end by a mirror.

Because of the type of material used (the type of crystal, gas or liquid) most of the light energy is being created at specifically selected frequencies. The energy which is at unwanted frequencies is allowed to escape through the sides of the tube, but the mirrors are angled so that the energy at the chosen frequencies is folded back in. The trapped light energy – or electro-magnetic radiation – bounces to and fro between the mirrors,

exciting more electrons to produce electro-magnetic energy and an intense light builds up.

The important thing is not just the emission of electro-magnetic energy by an electron being pumped up and falling back, but also the ability of the energy coming from one electron to cause another electron to make a transition and emit its own electro-magnetic radiation. In this way the light is amplified. In other words, what is happening is Light Amplification by Stimulated Emissions of Radiation.

If the two mirrors enclosing the tube were perfect, nothing more could happen, but the one at the business end is designed so that it will allow a proportion of the light to pass through it. This light is very intense, not only because it has been amplified, but also because it is all on the same frequency or wavelength (or, at least, very nearly so). Therefore, as all the waves are in step they reinforce each other. This light is said to be coherent or 'in phase'. What is more, the long narrow construction of the laser tube means that the beam of light hardly diverges outwards at all.

The laser has produced a light which is not only extremely bright, but also intensely powerful because it is in phase and travelling straight ahead in a narrow beam . To continue the analogy from the previous page, think of the photons leaving the laser as people. Whereas the battery torch gave out a random beam of 'people', the laser gives out an army – in step and marching in a straight line.

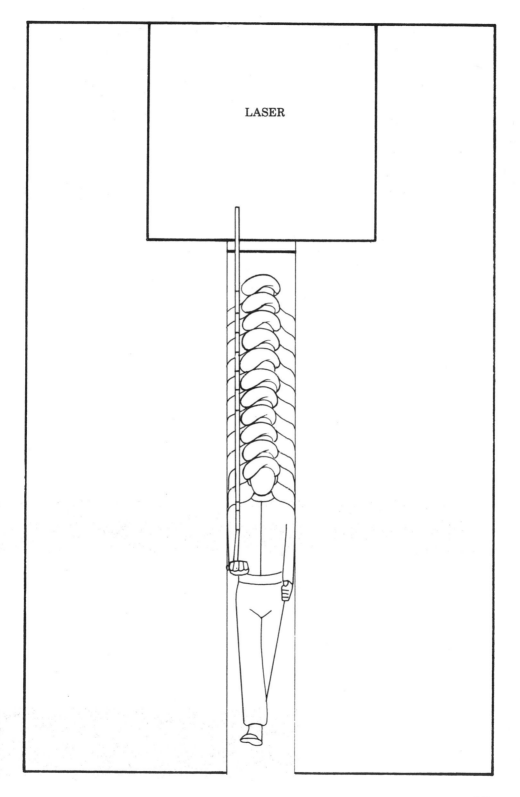

LASER

Before the laser there was – and still is – the maser (which was itself created as a result of research into radar). Maser is an acronym for Microwave Amplification by Stimulated Emission of Radiation. Microwaves vibrate on frequencies far higher than those of visible light. Among their practical uses to man are the high-speed heating effects of the micro-wave oven and the accurate running of a standard clock. It was as a result of work with micro-wave amplification that the possibilities of light amplification were discovered. The first known successful laser was built in 1960.

This first laser was – like most of those which followed – an amazingly simple device considering the highly dramatic effects it could produce.

Its heart was a single crystal of synthetic ruby, shaped like a rod, its ends smooth and flat. One end was completely silvered over to form an inwardly reflecting mirror, and the other end was partly silvered, to form a mirror which would reflect but would also allow some light to escape.

Its energy source was light – a powerful xenon flash-lamp, as used in strobe lights, enclosing the crystal and discharging energy in brief bursts, a few milliseconds apart.

The ruby is mostly made of aluminium oxide, but it contains

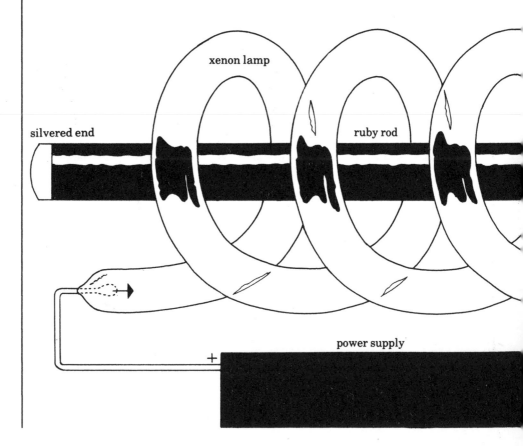

xenon lamp

silvered end

ruby rod

power supply

+

traces of chromium and it is the chromium atoms which become excited by the light energy and begin to shoot out photons of light themselves.

Because the photons fire off in all directions, some radiation is lost through the sides of the laser, but the photons which are travelling to and fro along the rod stimulate the atoms to produce more photons going the same way. These photons race to and fro, bouncing off the mirrors, creating more companions on their own wavelength, until they escape through the partly-silvered mirror as a highly intense pulse of laser light.

The whole process, from the switching on of the xenon lamp to the emission of the laser beam, is so rapid that it appears to an observer to be instantaneous.

Because the power source is a flash lamp, the beam from this type of laser is not continuous but comes out in a series of powerful pulses. But again, the speed of the operation is such that to an observer the thin, brilliant line of light appears to be continuous.

The ruby (and other crystal) lasers are still very much in use and in fact are the most powerful lasers of them all, albeit for very very short durations.

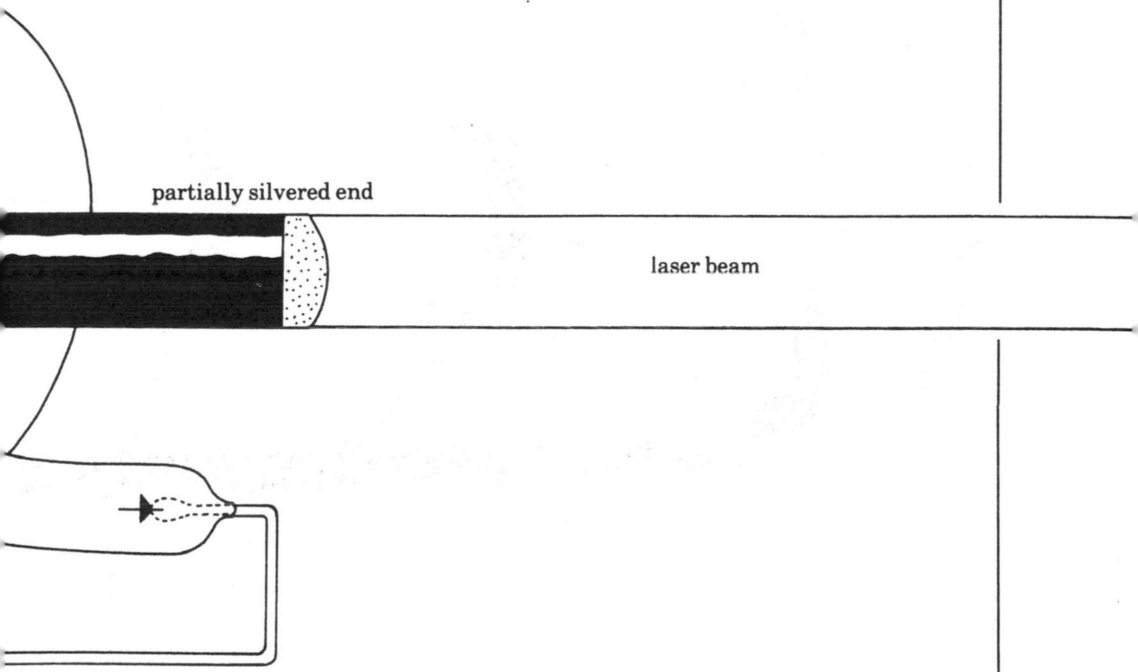

partially silvered end

laser beam

The first successful gas laser was built in 1961. The lasing medium was a mixture of helium and neon gas, and this type of laser – known as the HeNe – is the one most commonly in use at present.

The heart of this laser is a long narrow tube, usually about 0.1 cms in diameter and from 5 to 100 cms long, filled with the gas mixture, and with the usual pair of mirrors, one at each end, one fully reflecting, the other partially.

The energy input for a gas laser is almost always electrical, and the principal on which it operates is the same as it is for the crystal laser. The mixture of helium and neon is used because the atoms of the two gases have the ability to excite each other to higher energy levels, given a little help from an outside energy source. The electrical power, pumped into the laser tube, excites the helium and neon atoms and the general activity in the tube causes them to collide with each other, creating yet more excitement. The electrons of the affected atoms rise to higher orbits and then jump back, each firing off a photon of light as it does so. What seems like internal chaos results in the coherent and powerful beam of laser light.

The gas laser is actually less powerful than the ruby laser, but

fully reflecting mirror

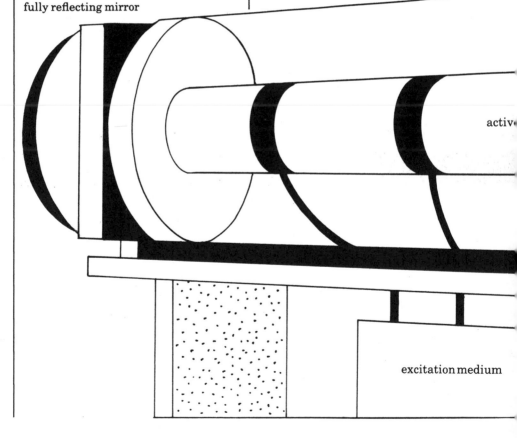

active

excitation medium

it has other advantages. Because the electrical input is steady (unlike the flashing lamp) the beam can flow continuously as well as in a pulse. Also, by its very nature, a gas has more possible energy states than a solid, so it has more possibilities of lasing. What is more, it does not have the problem of overheating that a crystal has. Lasing is only possible if the electrons make a transition to a higher energy state and *then* cool and fall back. A crystal cannot get rid of its own build-up of heat which means that it can overheat and distort. In theory,

if the crystal could be made to continue lasing when it had overheated it would distort so much that it would break. In fact that stage could never be reached because the electrons, being unable to drop back to lower energy levels, would no longer be capable of giving off energy and lasing would stop. A gas, however, can cool by its own currents, by a process of convection. This means that a gas laser can accept greater throughputs of energy and can be kept lasing for longer than its crystal predecessor.

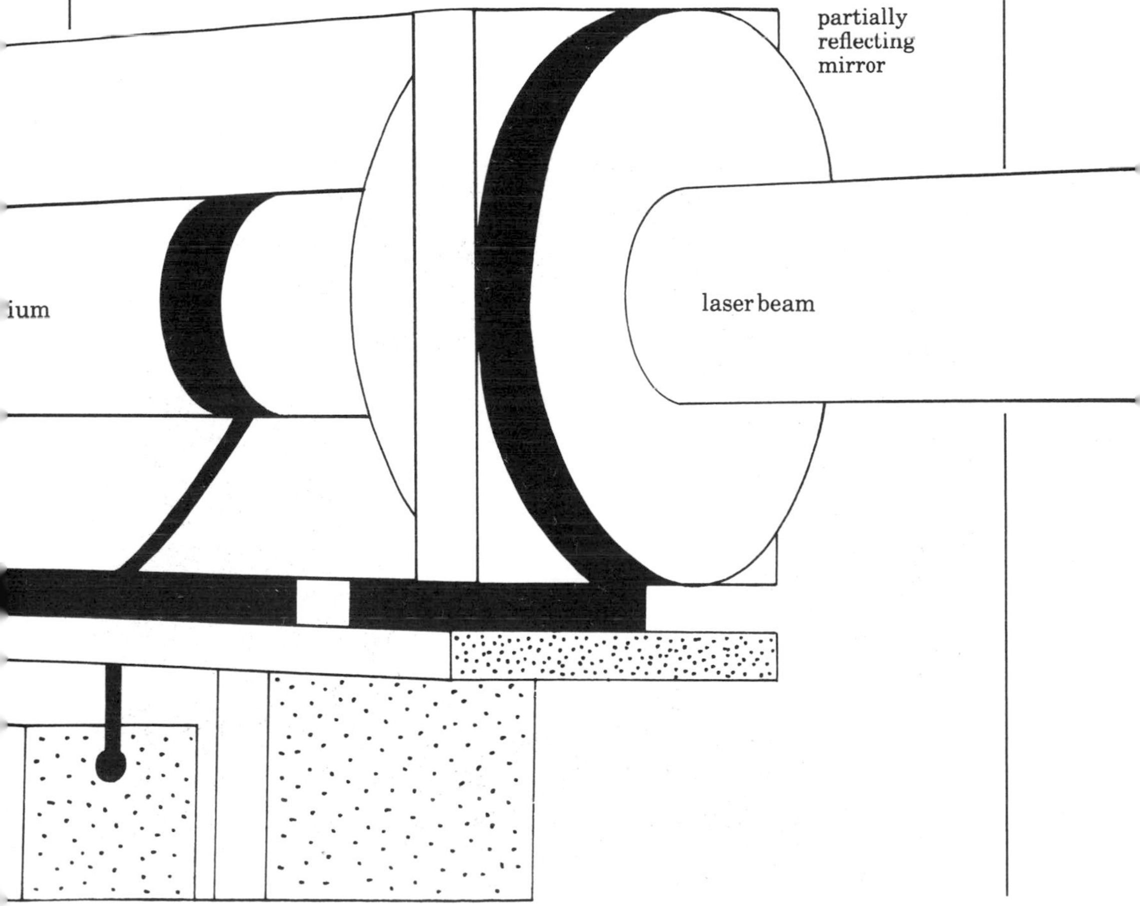

partially reflecting mirror

ium

laser beam

The semi-conductor laser can be very small and gives out a relatively low-powered laser beam. It works on the principle of a diode, which is a device designed to allow electrical current to flow in one direction only. There are several types of diode, but the one used for lasing is made from a semi-conducting material which has had some impurities artificially introduced and very precisely positioned.

As with the crystal and gas lasers, power is pumped in – in this case, electrical power. The resulting activities of the atoms and their electrons produce phased light waves along the line of impurities.

The laser beam that is emitted from most laser diodes is in the infra-red range of the spectrum.

In actual fact this type of laser does not give out a single beam but lots of little separate beams – each one a true, coherent, laser beam – which travel so closely together that the appearance and effect is of the more usual, single beam.

Below a magnifying glass shows all the separate beams of this laser.

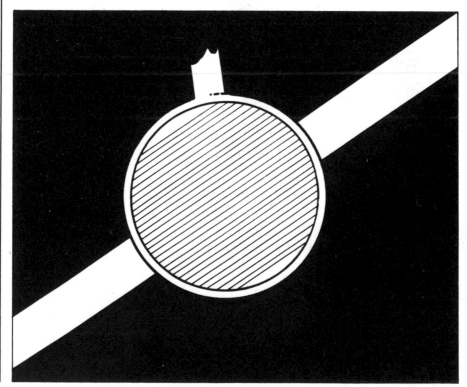

Organic dyes were relatively late discoveries among the laser-active materials, but they have proved invaluable. The dye can be used as a lasing medium in its solid, gas or liquid forms. The liquid or gases are preferable because they are not damaged by intense laser activity.

The power source for a dye laser is usually either light – as for the ruby or other crystal lasers – or else a gas laser.

The dye laser has two major advantages over its older relations. The organic dye itself is cheap and, more importantly, the dye laser has a broad spectral bandwidth – that is to say it can be finely tuned over a wide range of wavelengths. A single dye laser can range all the way from near-ultra-violet to near-infra-red. The dye itself determines the possible range. The structure of the laser – in particular the size of the resonator in which the lasing takes place and the inclusion of prisms and lenses at certain precise angles – makes it possible to achieve very fine tuning, and to retune at will.

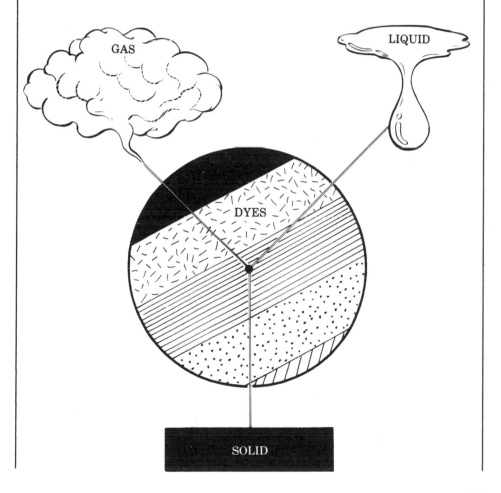

Lasers differ from each other in the power of the beam, in the area of the spectrum in which they operate – from ultra-violet to infra-red – and also in the manner in which the beam leaves the laser. This beam can be a continuous wave, pulsed or chopped.

The continuous wave, or steady beam of light, most usually comes from a gas laser.

A pulsed wave which is broken down into short but powerful bursts of energy – which are so close together that the beam appears to the eye to be continuous – usually comes from a ruby or other crystal laser, although most gas lasers can work in this manner if required.

A chopped beam – one in which the power goes up and down on a mean level but does not necessarily drop back to nothing (as a pulsed beam does) usually comes from a semi-conductor laser, although, again, gas lasers can use this mode.

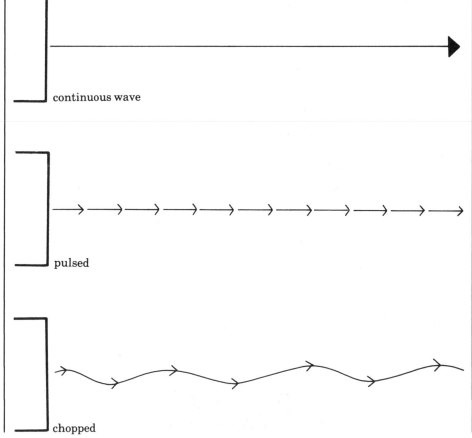

continuous wave

pulsed

chopped

The laser beam can be directed to its target in a number of ways. For example, the laser itself can be mounted on a small robot, or the beam can be bounced off a carefully angled mirror. But one of the most important methods is using a crystal placed in the path of the laser beam. If a small amount of electrical power is applied to certain crystals it can cause them to expand or contract. If a laser beam is passing through the crystal at the time, the change in the crystal's structure can bend it a little.

Some crystals become opaque when hit with electricity, which means they can be used to switch off the beam. This is particularly useful for industrial lasers. If the beam is switched on and off very fast, the amount of energy coming out in each burst is heightened which means that most of the damage (cutting or boring) is done by vibration (tearing) rather than by heat (melting) and so it is possible to work with inflammable materials.

The smaller lasers can be hand held.

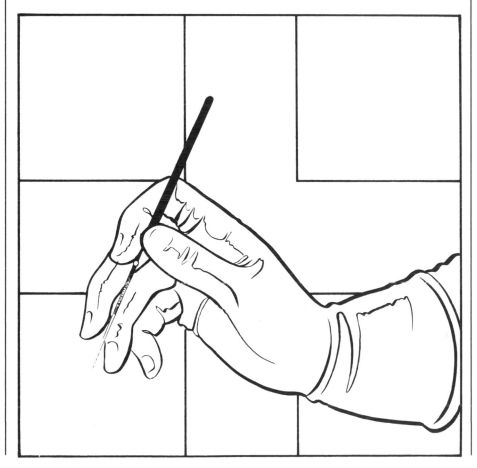

The laser is entirely dependant on the original input of energy – usually light or electrical – which makes the light amplification possible. But it is also dependant on getting exactly the right amount of energy for its size and capabilities. If it doesn't get enough, lasing stops or possibly doesn't even start in the first place. If it gets too much – or if too much energy is folded back in by the containing mirrors – the atoms reach saturation point and again lasing stops. The smaller the laser, the less tolerant it is to power-overload. (And the most obvious and instant defence against a laser used in warfare is to reflect its own power back into it so that it not only stops lasing but probably burns out as well.)

A phenomenal amount of energy is necessary to operate a laser, and this constitutes one of the main dangers if the energy in question is electrical. Even though protective shields and large warning notices (complete with diagram of high voltage bolt of electricity) may be

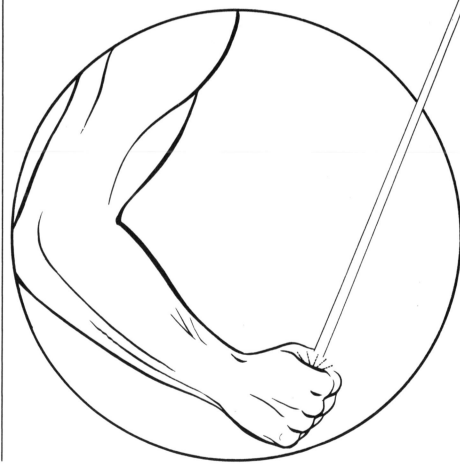

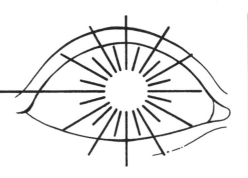

perfectly adequate, no amount of safety procedures can protect anyone from the results of carelessness. In practice, the danger to the laser operator is low – no greater than the danger to anyone using heavy machinery. It is those researching into new uses for lasers and more efficient methods of running them who are the most likely to take liberties with them, and to have highly charged accidents.

The most obvious danger comes from the cutting and burning action of the beam itself. It is not unknown for an operator to have a loosely swinging necktie unexpectedly severed, and a powerful beam would not take perceptibly longer to sever a finger or a hand.

However, the principal danger is to the eyes. The eye is obviously well-equipped to deal with ordinary light, but there is nothing ordinary about laser light. Even a very low-powered beam is only just safe to look at. The lasers used to create effects at pop concerts are of extremely dangerous power levels, which means the safety precautions are (or should be) very stringent. Those who work with industrial

lasers wear protective goggles with lenses so dark that only an acceptable amount of light can get through. But even these are only protection from a side-on view of the beam. A powerful laser focused directly on to the lens of a goggle would burn straight through it.

Again, the greatest danger is to those experimenting with and developing lasers. If a laser beam comes out from an unexpected direction, the eye turns instinctively towards the source of light before the brain has time to tell it not to. And if the laser beam is not focused directly on to the retina, where it can do most harm, the eye – being the efficient focusing mechanism that it is – adjusts the focus and by its own efficiency aids its own destruction.

But this grim scene is extraordinarily rare, and as an antidote to the feeling of revulsion it arouses it may be worth bearing in mind that a low-powered laser in the hands of a surgeon can effectively and painlessly restore the sight of someone suffering from a detached retina.

Because the modern Western world needs an enormous amount of power to keep its machines running, its lights lit, its interiors warmed and its lasers lasing, research into new – and preferably cheap – ways of producing power is vital to its survival in its present form. And one important source of power is nuclear energy.

Fusion and Fission

The production and use of nuclear energy is highly controversial. This is not only because of its potential in particularly unpleasant destructive weapons, but also because an accident with a nuclear reactor can be devastating. There is even the further difficulty of disposing of nuclear waste without harm to the environment. However, research continues into two methods of producing nuclear energy. These are fission – in which the energy is created when the atom is split in two – and fusion in which the joining, or fusing, of two atoms creates even more energy. The power of the so-called hydrogen bomb is achieved by fusion, whereas the power of the original atom bomb is achieved by fission.

Lasers are involved in this work because of their ability to create 'secondary stimulation' – in other words, to stimulate the atoms in certain gases or solids to behave in particular ways. Nuclear fission can already be inspired by laser light. Research into the powerful nuclear fusion is under way, and lasers may play a part here, too.

Fluorescence

Another effect of secondary stimulation is to create fluorescence in certain substances. Different substances radiate fluorescence at different frequencies which makes this particular ability of laser light useful in chemical analysis. The fluorescent wavelengths of specific chemical substances are known, so that when the analytical chemist directs a laser beam at an unknown substance he can measure the fluorescence it emits and come to a conclusion about its nature.

Feasibility Studies

Secondary stimulation is also used in feasibility studies on likely materials. Several gases, crystals and liquids are already known to be capable of lasing and are already used – the ruby or garnet crystal, the helium-neon gas mixture and the liquid dye are the best known. But there are certainly other substances with the same capabilities waiting to be discovered. In fact, since strong sunlight produces quite enough power to excite atoms into lasing, it is possible that a variety of rocks and crystals are lying around lasing in the sun at any given time!

But because there is no way for the photons of light to emerge as a beam, it is not possible to spot them in the wild, so to speak. Therefore, various likely substances are made capable of emitting their light and then tested under laboratory conditions, most often by stimulating them with laser light and observing their reaction.

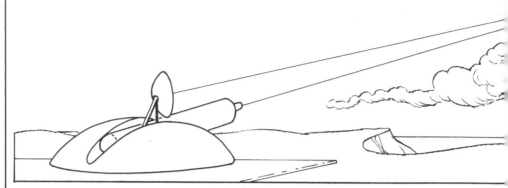

Oddly enough, the destructive power in a concentrated laser beam is not its most practical attribute when it comes to warfare. It is better at defence than attack, and better still (and this has been proved in modern warfare) as a secret light source to steal an advantage in night fighting.

It is true that the laser can be used as a kind of deathray, as it tends to be in the movies, but as a weapon of attack it has three serious limitations.

Firstly, it is not readily portable because it can only function if it is attached to a power source, and a laser vigorous enough to fell a missile needs an extremely large power supply.

Secondly, the narrow beam can only cause damage to an extremely small area of its target, which means that it must be aimed with perfect precision at a very vulnerable spot if it is to have any real effect. This is possible, but tricky, roughly on a par with shooting a charging rhinoceros through the eye.

And thirdly, it is not difficult to create a shield against laser light. All that is needed is a mirror or highly polished surface which can reflect the laser beam back on itself and cause sufficient power overload to stop lasing.

A laser operating in the infra-red frequencies can be focused through a lens which will spread the beam over a wide area. Anyone looking through optical equipment which is infra-red-sensitive can look at a scene that would otherwise be shrouded in darkness and see, quite clearly, the supposedly invisible night-time manoeuvres of an opposing army.

A defensive laser, permanently mounted in one spot, can be coupled to an automatic tracking radar which locks on to the vital electrical circuit guiding a missile, and to a computer which aims the laser and fires when it is on target.

As a defensive weapon, the laser certainly has possibilities. In fact, in the spring of 1983, President Reagan instructed USA scientists to being serious work on its development as a means of national defence.

Once a target is in sight, it is possible to use a laser to guide a conventional, remote-controlled bomb or missile to it. The laser illuminates the target and the missile homes in on the light energy reflecting back from its surface.

Fibre-optics is the name for the technique of carrying a communications laser beam along a thread-fine tube of spun glass, much as the electrical impulses which make up a telephone message are carried along overhead or underground cables. (In fact, fibre optics do have other applications, but it's their use with lasers that is relevant here.)

The reason for enclosing the beam in this way is that its strength, and the strength of the signal it carries, can be dispersed or generally interfered with if it is forced to pass through fog, mist or other atmospheric conditions that limit visibility. (This is a problem which does not, of course, occur in space; just as well for the development of laser inter-satellite communications since it would hardly be practical to string threads of spun glass between widely spaced pieces of extra-terrestrial technology.)

It has been possible for many years to transmit along radio waves, which are at the far end of the electro-magnetic spectrum (see diagram on pages 8 & 9). The signals are transmitted by modulating the wave – that is, changing the amplitude or frequency. But light can be modulated as readily as sound, most commonly by varying the brightness or by chopping the beam on and off, and the advantage of light waves is that they have a far greater carrying capacity than radio waves. Ordinary light is too weak and too chaotic, but the laser, with its ability to select one coherent wavelength which can travel great distances, produces an ideal light.

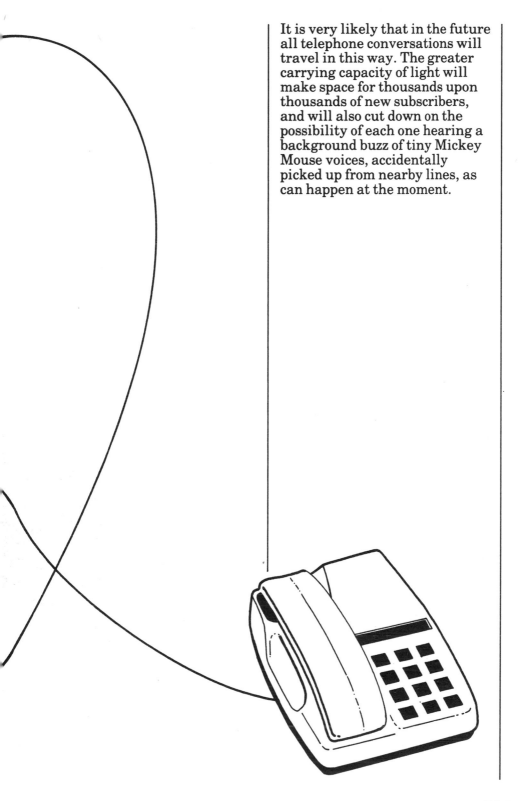

It is very likely that in the future all telephone conversations will travel in this way. The greater carrying capacity of light will make space for thousands upon thousands of new subscribers, and will also cut down on the possibility of each one hearing a background buzz of tiny Mickey Mouse voices, accidentally picked up from nearby lines, as can happen at the moment.

Lasers which give out visible light can be used for line-of-sight communications in situations in which fibre optics would not be practical.

Ordinary light has been used in this way for centuries, from beacon fires, through distress flares, to powerful lamps flashing morse code from ship to shore. But the laser has two main advantages over ordinary light. One is that it can operate in the infra-red part of the spectrum and so send messages which are invisible unless picked up through a specially constructed infra-red-sensitive viewer – particularly useful in military operations. The other is that the laser light has a very low divergence.

Ordinary light diverges outwards in all directions and eventually fades away, which limits its range. Coherent laser light diverges so little that it can cross immense distances without losing its clarity. This travelling ability has been successfully used to check the exact distance of the moon from the earth at any one time. The first astronauts to land on the moon set up a large, specially structured mirror. Astronomers in California can direct a beam of laser light at the mirror and then measure the time it takes to travel to the moon, strike the mirror, and reflect back again. Since the speed of light is known, this makes accurate measurement of the distance possible. No other man-made light would have been bright enough or coherent enough to make such a journey. Lasers truly illuminate the parts other lights cannot reach.

The laser light is not often called upon to cover such extreme distances. It can, for example, be used for line-of-sight communication between an oil rig and its land base.

One of the disadvantages of using laser light to communicate across the surface of the earth is that it can become absorbed in the atmosphere and so dimmed. Beyond the earth's surface, however, neither the atmosphere nor the problem exist, and laser light is ideal for satellite to satellite communication. Beyond the atmosphere, too, the beam can actually carry a signal, as a radio wave does, without the use of fibre optics.

It is the straightness of the laser beam that is so useful to the surveyor. It is like a long ruler of light that cannot sag or bend over long distances as a solid ruler or tape can. It makes precise alignment and precise measurement far easier than they were before.

The distance is actually measured by counting the pulses of light so that it is known how many pulses were needed to get from A to B. This is not possible with ordinary light because it is out of phase – it can only be done with coherent laser light. An easy way to understand why this

Distance can be measured by two surveyors, one at each end of the area to be measured. The first surveyor directs a laser beam at the second one, who reflects it back to him again. Sensitive equipment, usually attached to a computer, can assess the length of time it took the light to make the journey and so, knowing the speed of light, to calculate the distance covered.

should be so is to go back to the earlier analogy of the marching army. Think of the pulses of light as the footsteps of the army. If an army is marching in time it is perfectly possible to count the footsteps, left right, left right. But if instead of a well-drilled army you listen to a large crowd of commuters walking towards a station, all with different strides, all you will hear will be a blur of sound with no possibility of counting footfalls at all.

The straightness of the beam makes it ideal for aligning things, such as the two ends of a new bridge. A low-powered laser beam can be sent across the river from the chosen point on the first side, and the engineer on the other side knows that he can only see the point of light when he is directly in line with it. (The beam is only visible from the side if it is passing through atmospheric conditions, fog or smoke for example.)

When measuring heights in construction work, particularly if building on the side of a slope, there are often so many different levels involved – walls, ditches, steps, mounds, slanting ground – that there is no obvious 'ground level' to use as zero. A low-powered laser set on a tripod at a central point and at an agreed zero level can be revolved, so giving engineers on all sides of the work an accurate point from which to measure other heights and make calculations.

The tracking abilities of the laser are of obvious value to the military, who can illuminate a dark battlefield with invisible light or use a laser in the infrared area of the spectrum to lock on to a heat source and so seek out hidden personnel. But there are peaceful applications of these facilities, as well.

Some colours travel through water with less absorption than others and coloured laser beams – often deliberately diffused through a lens – are used for undersea exploration, searching out wrecks, discovering mineral possibilities and practical places to lay cables.

When an airfield is troubled by low cloud, it is possible to bounce the beam of a ground-based laser off the cloud cover and, by measuring the pulses on its return, to calculate not only how high above the airfield the cloud is lying but also how dense it is; valuable information for those who have to assess when the airfield is likely to be operational again.

To return to a largely military use, the coherent laser light can be used for submarine detection from the surface. The light bounces back far more strongly from the metal casing of the submarine than from the empty sea, or from rocks or shoals of fish.

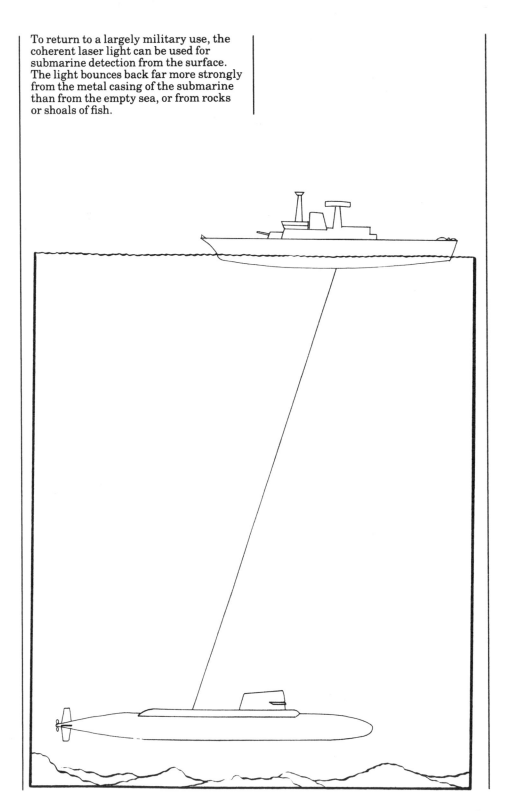

The lasers used in industry generally operate in the infrared part of the spectrum. It is their intense heat that makes them invaluable for cutting, drilling and welding at very high speeds and with very high degrees of precision. Because the laser is adjustable, and the power of the beam can be regulated according to the material to be attended to, lasers can be used for cutting through something as soft as several thicknesses of cloth or something as hard as industrial diamond.

Cutting

When the beam strikes its target, perhaps an area the size of a pin-point on a steel plate, it heats it almost instantaneously to a temperature as high as the surface of the sun itself. Because there is no time for the heat to disperse, a minor explosion is caused, which blasts microscopic portions of steel outwards in a miniscule cloud of smoke, leaving a hole. It is this small explosion which is seen by the naked eye as a white flash with accompanying sparks, and which is heard as a hard, sharp, metallic spitting sound.

Boring

Precision-boring of a diamond is a particularly useful attribute. Copper wire, and the fine glass tubes used in fibre optics, are made by drawing a piece of copper or glass, at high temperature, through a series of dies, each one with a smaller hole, until it passes out of the last hole as a thread of the exact

the advent of the laser, this problem has ceased to exist. In fact it is this advance in the manufacture of precision diamond dies that has made possible the huge advances in fibre optics that have taken place during the last few years.

Welding

A high-powered beam cuts – a beam of lower power can partially melt metal without actually cutting through it and so can be used for welding. In particular, it is valuable for

fineness needed. The material used for these dies must be extremely hard, otherwise the pressure caused by the operation gradually enlarges the holes. The problem is that extremely hard material, such as diamond, is extraordinarily difficult to bore through by conventional means, especially when the hole must be of a very small and also very precise diameter. But with

welding very tiny pieces of metal that traditional welding techniques would be too clumsy to deal with. In fact, lasers have contributed greatly to the development of the micro-electronics industry because of their ability to produce a minute point of intense heat – sometimes switched on for as little as one five-thousandth of a second – which can weld hair-fine wires onto components, or cut five hundred microchips from a wafer of silicon no more than 5 centimetres in diameter.

The Surgical Laser

When a low-powered, continuous-wave laser beam, operating in an infra-red frequency, is brought into contact with human tissue, its heat has a precision-burning effect that can have exceedingly useful results. The heat has both a welding and a cutting effect. It can seal off blood vessels and it can destroy unwanted tissue, or even bone, with a seemingly super-human accuracy that is, nevertheless, entirely dependent on the skill of its operator.

Heals as it Seals as it Cuts

Because the beam seals off blood vessels even as it cuts, the laser is able to overcome one of the most basic of surgical problems. That is, the fact that cutting a hole in a patient and digging unwanted bits out tends to cause excessive bleeding (which has a bad effect on the patient's health and also on the surgeon's ability to see what he's doing). People who have had their tonsils removed by laser have lost very little blood, felt no pain and hardly had any swelling of the throat to contend with afterwards. Liver surgery, in particular, stands to benefit, since the liver is virtually made of blood and tends to pour it out with unnerving speed at the first incision.

The beam's ability to cut off blood vessels and cause coagulation means that it can seal a bleeding stomach ulcer which could otherwise take months to heal. It also makes an invaluable tool for welding a detached retina back into place and so restoring sight.

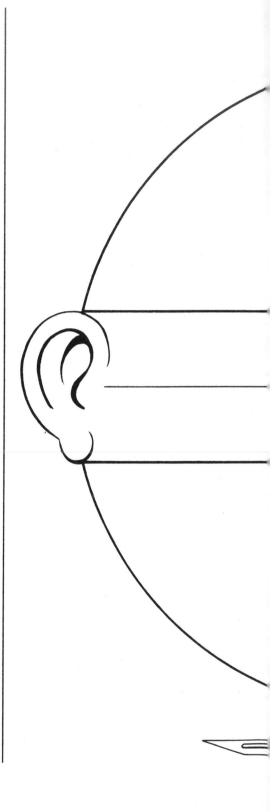

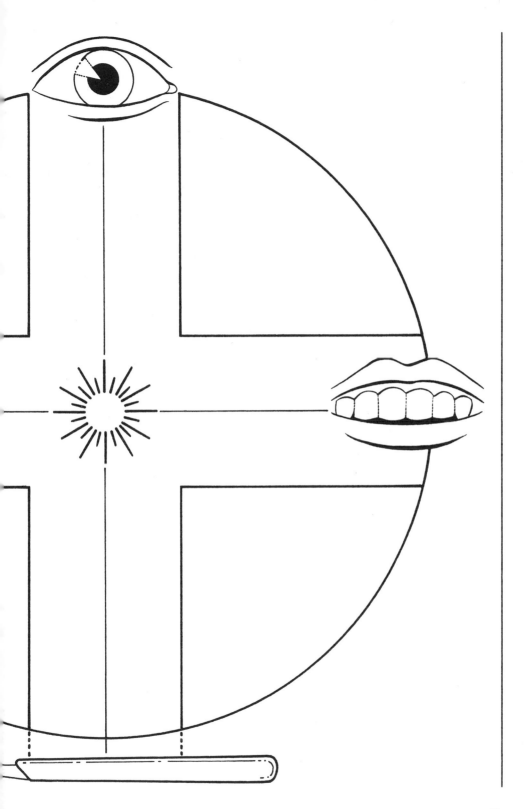

Restoring the Sight

In the case of eye surgery, a beam of ordinary light is often used to direct the laser beam to its target. The relationship between the two beams is measured with finer than hairline accuracy so that when the ordinary beam is directed at the retina in the good eye, the laser beam is known to point precisely at the damaged retina in the other eye. Only when the machine is perfectly aligned is the laser switched on, usually for a period of time which can only be measured in milliseconds.

Repairing the Ear

Its fine precision and cutting power mean that it can be used for the most delicate of ear operations. Hearing is only possible if sound waves vibrate on an extremely delicate construction of minute bones in the ear. If one of these is misplaced or if, instead of pivoting freely as they should, the bones solidify together, the inevitable result is deafness. An accurately directed laser beam can remove the offending bone without touching its neighbours.

Drilling the Teeth

Lasers are not yet very popular as dentist's drills. Since the cutting action is so rapid, and since it is not yet possible to measure the depth of decay with absolute accuracy, there still remains the hideous possibility that the dentist might not know when to stop. Also, the smell is terrible. However, it is already possible to direct a laser beam into a hole drilled through the enamel in the ordinary way and painlessly to seal off a nerve. Experiments are also being carried out with a new filling material which can be painted into the hole and then fused to the tooth by laser light.

removing tattoos

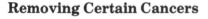

Removing Certain Cancers

Some cancers can be treated by laser beam, but with most the problem is that the malignant cells are not always very distinct from the healthy cells. In an imprecise situation, the laser's strength and precision can actually be a liability. But skin cancers, and small superficial tumours, can be removed very effectively.

Removing Tattoos

Lasers which operate within ultra-violet frequencies can be used to change the pigmentation of the skin – in essentially the same way as the ultra-violet in sunlight causes skin tanning and, as a secondary action, bleaching. This means that a laser – usually a small, hand-held one that looks deceptively like an ordinary torch – can obliterate embarrassing tattoos. It also works well on birthmarks; although it can't remove them completely it can cause them to fade and fade until they are barely noticeable.

Remaining Problems

Like any new technique, this one is not without its own problems. Two of the most obvious of these are the availability or otherwise of the expensive equipment, and the availability of surgeons trained in its use. A mistake – leaving the beam on for too long, perhaps, or misdirecting it by the smallest possible degree – could be devastating. In fact in the most delicate operations, fine glass fibres are used to carry the beam in order to direct it with ultimate precision.

Some of the small-scale, low-key uses of laser beams are just as fascinating as their more obviously dramatic applications. Because it is still tempting to think of the laser solely as lethal light, it seems more than a little eccentric to fit one to a domestic appliance – a record player, for example. But it is being done – although as it is still at a very expensive stage of its development it is not possible to say whether or not every record player will one day be equipped in this way.

What happens is that a small, low-powered gas laser, which consists of a glass tube full of helium and neon gas with an electrical filament (similar to the one in a light bulb) to produce the power, is designed to take the place of the arm of the player, so that the fine beam can take the place of the stylus. (The power of the beam is too low for it to slice through the disc.)

Glass tube filled with helium and neon gas

Because it is finer than any solid stylus could ever be, it can play far narrower tracks than a solid stylus could ever play. This means that more music can be packed on to one disc. Also the beam of light is never blunted, as an ordinary stylus will eventually be. However, it does wear out in the end. The HeNe gas in the tube burns itself out as it is used and the mini-laser has to be renewed.

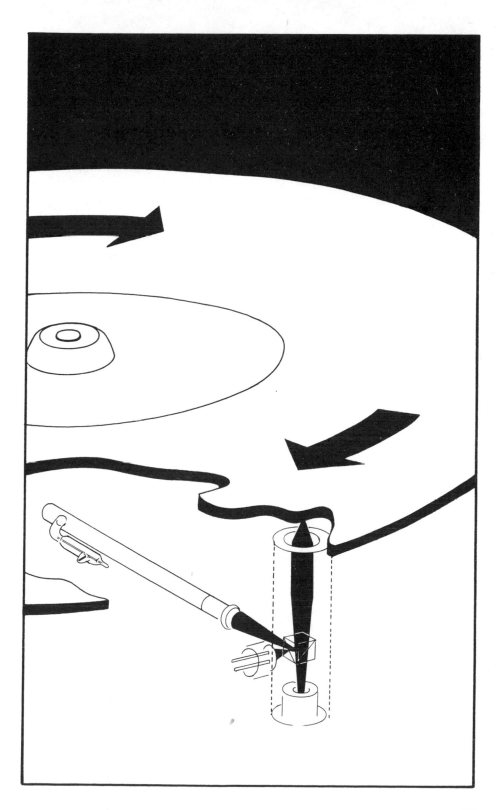

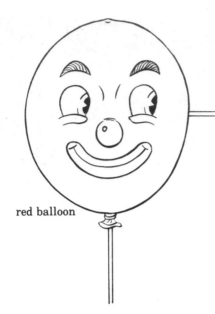

red balloon

Various simple experiments, almost amounting to games, are possible with the smaller lasers. For example, an object that is seen as red is actually one which is giving off light in the red wavelength and absorbing the rest. So if a red laser beam is directed on to a red balloon, the balloon can re-transmit the light with no trouble and no ill consequences. But direct a blue laser beam at the red balloon (or a red beam at a blue balloon) and the light will be absorbed, not re-transmitted, the rubber will overheat and the balloon will pop.

It is also possible to focus a laser beam so precisely that the molecules of oxygen and nitrogen in the tiny portion of air it is affecting, heat up and push away their surrounding molecules with a pop that can be heard quite clearly.

For most people, still, lasers mean light shows and special effects at the more lavishly designed pop concerts. These lasers need a surface on which to project their strange effects, and this is usually a screen. If a static screen seems too limiting in the context of a show, then smoke is used – literally a smoke screen, on which the patterns can be formed.

It seems incredible that such a powerful tool can be used in such a light-hearted manner, without

blue balloon

46

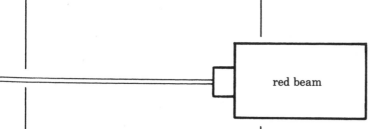

red beam

putting performers and audience at risk, and it is certainly true that stringent safety rules have to be observed. As was pointed out earlier, it is not the cutting and burning power of the beam that constitutes the principal source of danger. It is the potentially blinding brilliance of the light. Designers of light shows must ensure that the beams themselves can never be directly seen, only their reflections on the screen, and that the whole set-up is arranged so that the reflections themselves are many times below the levels of brilliance considered to be safe. Rules which cover these hazards can be laid down and followed in every situation – but whenever a new set-up is being laid out, or an existing one moved to a new location, extra care is necessary. Ironically, blindly following the rules is not enough. It is

essential, for example, to carry out extremely careful checks on the possibilities of unplanned reflections, from unnoticed polished surfaces, which could be almost as dangerous as the original beam.

Obviously, different lasers are capable of producing different effects and different colours, and the uses of lenses and filters can spread, direct and change the light in an almost unlimited range of possibilities.

The lasers are not only used as a background to a performance of music or dance – they are frequently manipulated to create their own show – and they can even be linked to the speakers so that the light patterns relate directly to the sound patterns of the music and singing, to produce an electronic version of Walt Disney's painstakingly animated film *Fantasia*.

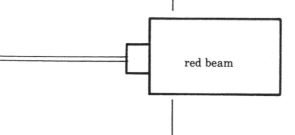

red beam

One of the oddest spin-offs of laser technology is holography, the science of creating holograms. A hologram is an image on a photographic plate, usually of glass, which, when illuminated at precisely the right angle, is seen as three-dimensional.

Look at the hologram on the front of this book, preferably in daylight. Slant the book to left and right and, when the light strikes the hologram at the correct angle, you will see that the spiral looks rounded and three dimensional. Tip the book slightly towards you and you seem to see down inside the top of the spiral. (At an exhibition of holograms the viewers can be seen weaving and swooping like charmed snakes in their efforts to get the full effect.)

Holography is a specialist science, but it is perfectly possible for non-specialists to understand the basic split-beam technique, and to understand that two things make it possible – one is the fact that the light waves reflecting from any object can be recorded photographically, and the other is that laser light is coherent, with all its waves in step.

The plate is developed using a technique very similar to a normal photographic technique – but whereas the picture can be seen on a photographic negative, the hologram makes no sense to the naked eye.

For the hologram to be seen it must be illuminated by a very bright light which strikes it at exactly the angle at which the reference beam originally struck the plate. The viewing light need not be a laser, an ordinary spotlight will do. When the viewing light strikes the plate it reflects back in the form of the disturbed light waves recorded upon it. So the image of the subject is recreated.

At this stage the hologram appears as though behind the plate – as though you were looking through the plate at the subject. However, it is possible to make a hologram of the hologram – using the same technique as that just described, with the original hologram taking the place of the subject. If this second hologram is illuminated, the three-dimensional image of the subject seems to appear in front of the plate – between plate and viewer.

It is possible to record several images on a single plate by changing the angle of the reference beam for each one. If the plate is then turned through the beam of the viewing light, the images will appear and vanish in turn. This makes some fascinating effects available to those who create holograms for exhibition as works of art.

The diagram opposite shows how the hologram on the cover was created.

The coherent light of a laser beam (1) is directed by a mirror (2) into a device known as a beam-splitter (3) which does just that – splits the beam in two. One of these is known as the reference beam (4) which is directed by another steering mirror (5) through an expanding lens (6). This lens spreads the light on to a holographic plate at a precise angle (7). Simultaneously the other beam, known as the object beam (8) is directed by a steering mirror (9) through an expanding lens (10). This lens spreads the light all over the subject (11) which reflects it.

The light waves reflected from the subject meet the light waves from the reference beam (12) on the surface of the plate, which is coated with minute light-sensitive crystals. Because both light sources originated in the same laser they are of exactly the same wavelength. They are therefore able to coincide and create what is known as an interference wavefront. This interference wavefront records the shape and depth of the subject on the holographic plate.

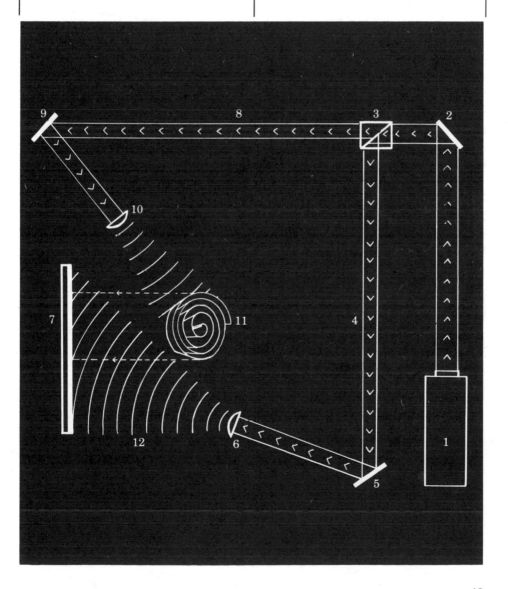

When the technique of holography was first invented, it had no obvious uses. It was, literally, a trick of the light, an optical illusion, a 'fun' by-product of the serious business of producing lasers for the industrialists, the surgeons and the military. It was actually described in its early days as a 'solution looking for a problem'.

However, a few practical problems to which it is a very useful solution presented themselves quite quickly, and the back-to-front research (to find the questions to the answer) continues.

Improving Fuzzy Photographs

One of the earliest uses was the clarification of the pictures taken by the first astronauts on the moon. Some of these were out of focus, but it proved possible to make a hologram of each and, by adjusting the angle of the subsequent beam of viewing light, to refocus the image – and then to take a new photograph of the in-focus hologram. A long-winded but effective method of producing sharp prints which could be examined in detail by the various experts wanting to look closely at the moon's surface.

Spotting Wobble

Another use was discovered accidentally when it was found that, for a successful hologram to be taken with a continuous-beam laser, the subject had to be perfectly still. (Only pulsed lasers can cope with moving targets, freezing each moment as a high-speed camera does.) Therefore a hologram – or, in a

sense, a failed hologram – is able to show if an apparently stationary object is actually vibrating or wobbling.

high speeds – undetectable by any other means – could result in disastrous stress. What is more, by using a pulsed laser to take two holograms on the same plate, one on top of the other, it is possible to determine how much the object is vibrating, and how fast, by measuring how much the second picture is out of register with the first.

Observing Micro-Organisms

Holography can assist scientists engaged in the study of minute organisms, which can only be seen through a microscope. When microscopic magnification is very great, it is only possible to bring a tiny area into focus at any one time. This makes it extraordinarily difficult to examine living organisms which can never keep still for a second. A photograph taken at high speed can capture the creatures, but it is not always much use because of the distortion that is inevitable in a flat picture of something which is three-dimensional. But a hologram taken with a pulsed laser can freeze the organism while still showing its every curve and detail for the scientist to examine at leisure.

Analysing Particles

It has also proved extremely useful to those engaged in particle analysis – that is, investigating the number and size of the minute particles floating in fogs, in the mist given off by aerosols, and so on. In this case, too, it can freeze an image of the sample without either disturbing it or flattening it.

This has proved invaluable in strain analysis on such things as aircraft and engine components in which minute vibrations at

Superior Photographs

The hologram is actually a very superior photograph so it can, anyway in theory, do whatever a photograph can, but better. It is, for example, capable of – wait for it – 'high-resolution imagery through aberrating media'. That is to say, it can produce a clear image of something through distorting glass or even through water. This facility already has scientific uses and may one day be sufficiently widely available to benefit those who favour underwater holiday snaps.

Recognising Patterns

Another possibility is to use the technique to speed up the recognition of specific patterns – to check fingerprints or post codes, for example.

Each set of fingerprints would be stored in police records in the form of a hologram. The finger-prints found at the scene of a crime would be recorded in the same way. Then the normal procedure for producing a hologram would be reversed.

That is, instead of the light waves which form the picture of the object being brought out of the hologram by the reference beam, the reference beam would be brought out by the object beam. Therefore, only when two identical holograms of prints met would the reference beam be activated. With a properly designed electronic set-up this could all happen very fast – but there still remains one snag which is, so far, insurmountable. The technique is so precise that it cannot cope with smudged fingerprints (or, indeed, with variable handwriting of post codes!).

3-D T-V

The idea of a television picture in three dimensions that hovers just in front of the screen, as though a theatre of ghosts had chosen to present a performance in the living room, is an exciting one, which would open up all sorts of possibilities for directors – most obviously in the areas of horror or sci-fi. At the moment huge problems stand in the way – not least the difficulties of eliminating the dangers to actors and of raising sufficient funds for research and trials. But theoretically it is a possibility – and lasers themselves were once no more than that.

Light Minds

Already optical memories exist which can store large quantities of the binary data used by computers. This kind of computer memory is made up of a series of holograms whose images consist of light or dark spots in which the information is encoded. The computer is designed with the ability to locate and illuminate the relevant hologram and to 'read' the information electronically. This 'light' memory has proved to be able to contain more information in a given space than older electronic memories, and to be far less likely to be upset by tiny scratches or dust specks. At the moment, this is probably holography's greatest area of growth and practical usefulness.

Holography is still a very young science, whose possibilities have by no means been fully explored. But it clearly has a future – and not just as the optical conjuring trick it was at first thought to be.